J
621.319 Lillegard, Dee
LI
   I can be an
   electrician

$12.60

AG 06 DATE		
JUL 23 '92 AG 11 '08		
OC 1 '92 AG 14 '08		
OC 29 '92 NO 1 1 12		
AUG 11 '90 MY 7 2 17		
FEB 06 '95 JE 2 0 17		
JUN 14 '95 MR 1 7 '22		
SEP 01 '95		
JUL 1 0 98		
FE 15 99		
AY 16 0N		
JY 09 '01		

© THE BAKER & TAYLOR CO.

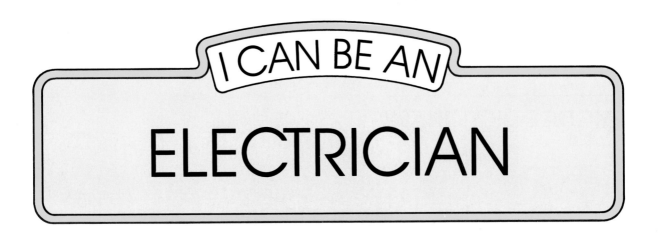

ELECTRICIAN

By Dee Lillegard and Wayne Stoker

Prepared under the direction of Robert Hillerich, Ph.D.

CHILDRENS PRESS ®

CHICAGO

Library of Congress Cataloging in Publication Data
Lillegard, Dee.
    I can be an electrician.
    (I can be)
    Includes index.
    Summary: Examines how electricians make, control,
and work with electricity and highlights the education
and training necessary for the field.
    1. Electricians—Vocational guidance—Juvenile litera-
ture. [1. Electricians—Vocational guidance.
2. Vocational guidance. 3. Occupations] I. Stoker,
Wayne. II. Title. III. Series: I can be.
TK159.L55 1986   621.319'24'023   86-9657
ISBN 0-516-01896-5

PICTURE DICTIONARY

lightning

electrician

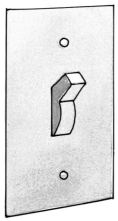

electrical
switch

battery

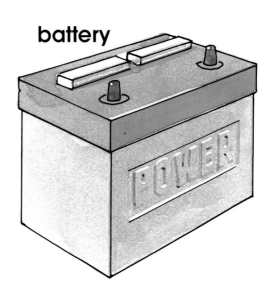

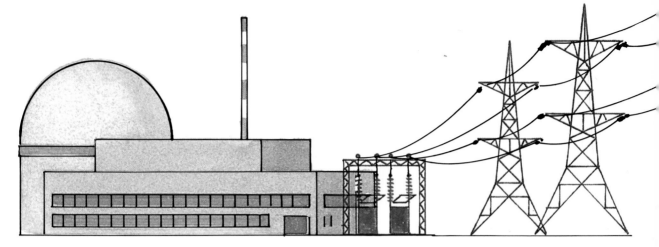

power plant

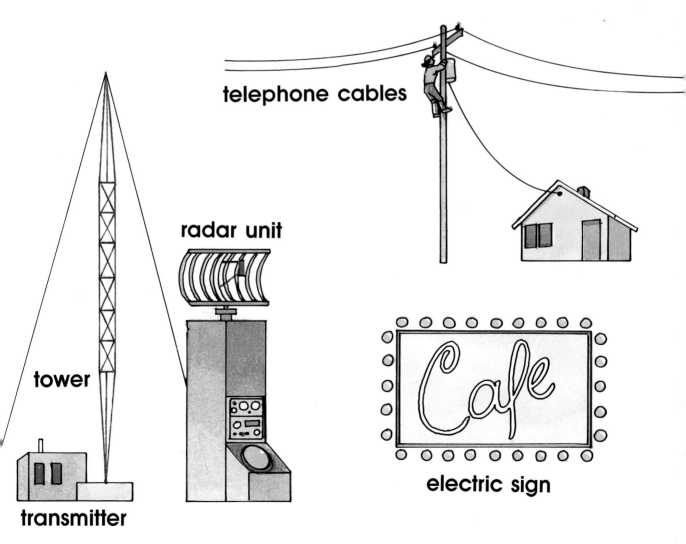

telephone cables

radar unit

tower

transmitter

electric sign

Cafe

Some people say electricians work magic. Electricians really work with a powerful form of energy. It is called electricity.

electrician

Electricity makes our toasters and televisions, our stoves and refrigerators work. Without electricity, cars would not start and airplanes could not fly.

Opposite page: An electrician working on high tension wires

Lightning is the most powerful form of electricity on earth. Benjamin Franklin (left) setting up an experiment with electricity using a kite

lightning

Over two hundred years ago, Benjamin Franklin used a kite in a thunderstorm to show that lightning is a kind of electricity. He was lucky that the bolt of lightning that struck the

Electricity gives us many more opportunities for work and play than we would have without it.

kite did not harm him.

Lightning is the most powerful form of electricity on earth. We cannot control lightning. But electricians can make—and control— other kinds of electricity.

A classroom experiment with electricity. As the boy cranks the generator, the bulb lights up.

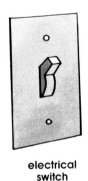

electrical
switch

Electricity flows in much the same way that water does. If you turn on a faucet, water will flow out. When you turn on an electrical switch, an electric current flows through.

But you cannot see or taste electricity. And it will not form a puddle. Electricity is a very special kind of energy. Electricians know this.

Young electricians may start out with home experiments. But to be safe, they never experiment with the electric current in their homes. That would be dangerous.

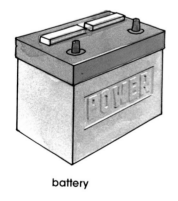

battery

There are many safe ways to experiment with electricity. You can produce electricity with a battery you make yourself. You can also make electricity with a magnet and a coil of wire.

An electrician training his apprentice

Electricians must first be apprentices. Apprentices work for an experienced electrician who gives them on-the-job training.

11

Students in a
trade school
learning electronics

Apprentices also take classes to learn mathematics and the science of electricity. They study electric currents and electric circuits. They learn how to measure electricity in order to send it from place to place.

13

Because cars have an electrical system, car mechanics need to know a lot about electricity.

Electricians must also understand electric batteries, motors, wiring, and lights. They must learn how to understand architects' drawings of new buildings so they can install the electric wiring properly.

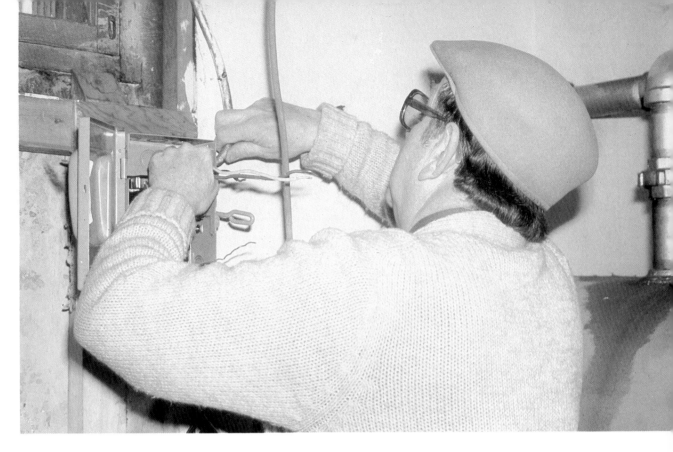

Electricians installing electrical systems in new buildings

An electrician's work can be very dangerous. These power lines carry 345,000 volts of electricity.

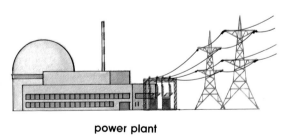

power plant

There are many kinds of electricians. Some work in power plants. These plants produce the electrical energy that flows into our homes and factories.

When working in construction areas, electricians wear hard hats for safety.

Others work in the
construction industry.
They install electrical
wiring in new homes
and office buildings.
This wiring carries the
electricity generated by

A power plant in Boston, Massachusetts

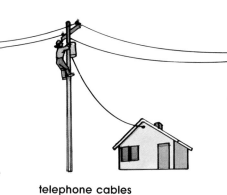

telephone cables

power plants. The electricity makes lights, heaters, fans, and elevators work.

Many electricians work in communications industries. Some work with telephone cables.

These are some of the jobs electricians do to make our telephones work.

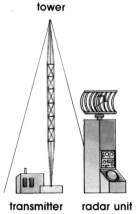

tower

transmitter   radar unit

Others work in radio or television. They may work on towers, on radar units, or on

transmitters. They may even work on electric signs that advertise products.

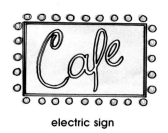

electric sign

Left: Repairing street lights. Above: Fixing phone lines that have been damaged during a storm

Some electricians are maintenance workers. They service and repair electrical equipment. They keep our power plants, factories, and

These electricians are maintenance workers for large office buildings.

hospitals working—and
our schools, offices, and
homes, too.

Electricians often work
"straight days," going to
work in the morning
and coming home in

the evening. Or they may work in shifts. This way, their work hours depend on the kinds of jobs they are doing.

Some electricians work in the same place every day. Others go from one place to another.

Electricians need to
have good eyesight,
hearing, and sense of
smell. They must be
healthy and alert. And
they must be careful
about their work.

y is powerful.

dangerous.

s work with

elec city can lead to electric shock or fire. You know that careless bicycle-riding can be dangerous. Electricians, like bike-riders, learn to pay attention—all the time.

We need electricity.
Without it, our cities
could not survive. Our
lights would go out. Our
factories would stop.
Our lives would be very
hard.

Electricians make electricity go wherever people need it. This might seem like magic. But electricians know it is not. It is their kind of work, and they are proud of it.

WORDS YOU SHOULD KNOW

apprentice (up • PREN • tis)—someone who is being trained on the job by a skilled worker

architect (ARK • ih • tekt)—a person whose job is to draw plans for new buildings

battery (BAT • er • ee)—a device that produces an electric current

cable (KAY • buhl)—a rope of electrical wires coiled together

circuit (SUR • kit)—the path that an electric current follows

electric current (ee • LEK • tric KUR • ent)—a flow of electricity

electricity (ee • lek • TRIH • sih • tee)—a form of energy that consists of the flow of tiny particles called electrons

lightning (LITE • ning)—a flash of light produced by electricity between two clouds or between a cloud and the earth

maintenance (MANE • teh • nents)—repairing something or keeping it in good condition

power plant—a station for generating electric power

radar (RAY • dahr)—a device that uses radio waves to find or watch an object

shift —a regular time period for work

shock—a serious jolt to the nerves and muscles caused by the body coming in direct contact with an electric current

transmitter (trans • MITT • er)—the device at a radio or television station that sends programs out of the station and into the individual sets

INDEX

PHOTO CREDITS

ABOUT THE AUTHORS

Dee Lillegard (born Deanna Quintel) is the author of over two hundred published stories, poems, and puzzles for children, plus *Word Skills,* a series of high-interest grammar worktexts, and *September to September, Poems for All Year Round,* a teacher resource. Ms. Lillegard has also worked as a children's book editor and teaches writing for children in the San Francisco Bay Area. She is a native Californian.

Wayne McMurray Stoker is a Culinary Arts instructor at Laney Community College in Oakland, California. A tradesman at heart, he has been involved in the building and manufacturing trades all his life and cannot resist exploring what makes things work. Having spent his early childhood in the rural South, where storytelling was a natural pastime, Mr. Stoker finds writing for children to be an enjoyable extension of his widely varied experience.